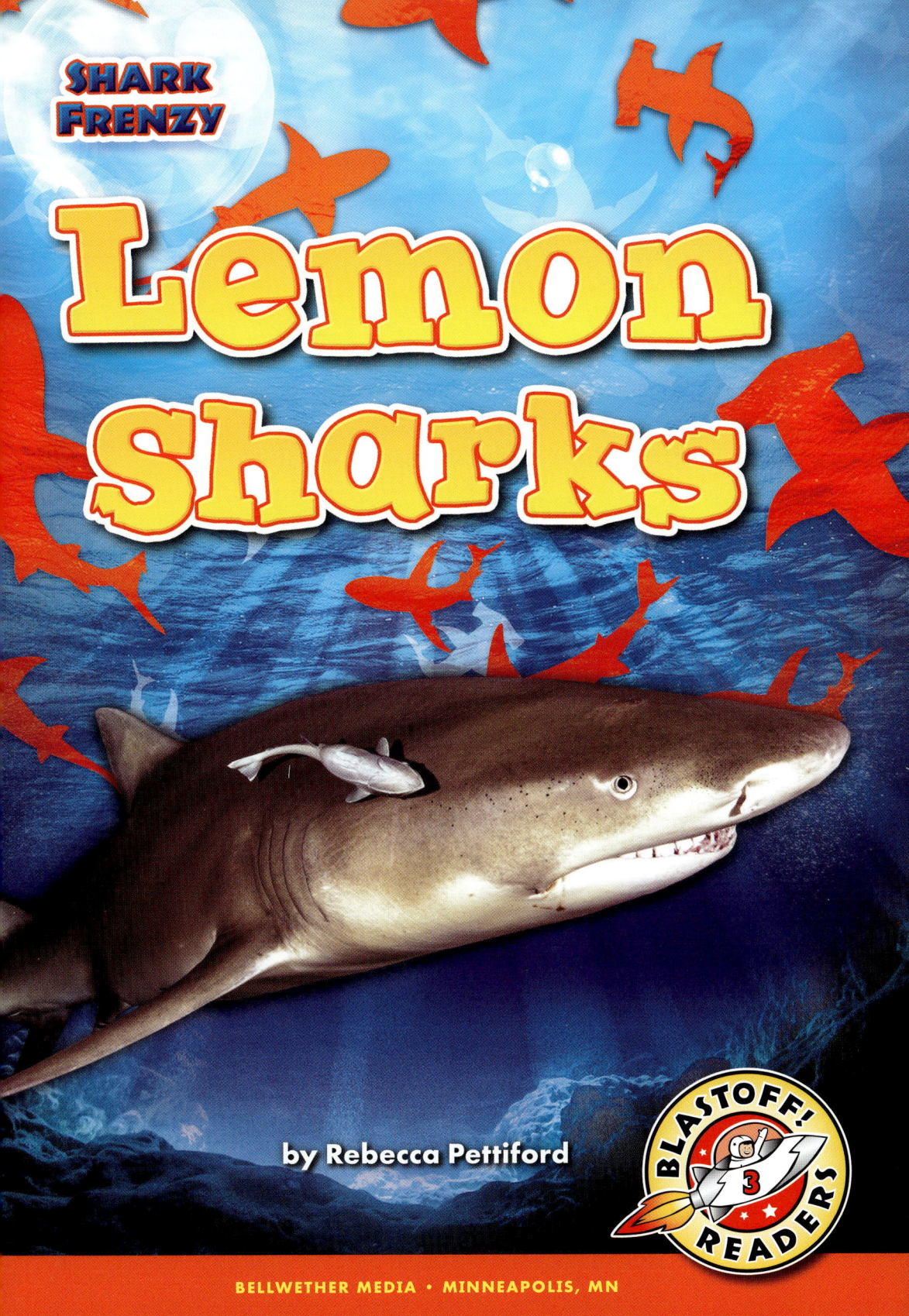

Blastoff! Readers are carefully developed by literacy experts to build reading stamina and move students toward fluency by combining standards-based content with developmentally appropriate text.

LEVELS

Level 1 provides the most support through repetition of high-frequency words, light text, predictable sentence patterns, and strong visual support.

Level 2 offers early readers a bit more challenge through varied sentences, increased text load, and text-supportive special features.

Level 3 advances early-fluent readers toward fluency through increased text load, less reliance on photos, advancing concepts, longer sentences, and more complex special features.

★ **Blastoff! Universe**

Reading Level

Grade K

Grades 1–3

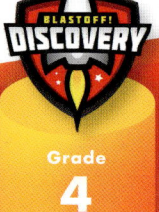
Grade 4

This edition first published in 2021 by Bellwether Media, Inc.

No part of this publication may be reproduced in whole or in part without written permission of the publisher. For information regarding permission, write to Bellwether Media, Inc., Attention: Permissions Department, 6012 Blue Circle Drive, Minnetonka, MN 55343.

Library of Congress Cataloging-in-Publication Data

Names: Pettiford, Rebecca, author.
Title: Lemon sharks / by Rebecca Pettiford.
Description: Minneapolis, MN : Bellwether Media, 2021. | Series: Blastoff! readers: Shark frenzy | Includes bibliographical references and index. | Audience: Ages 5-8 | Audience: Grades 2-3 | Summary: "Simple text and full-color photography introduce beginning readers to lemon sharks. Developed by literacy experts for students in kindergarten through third grade"– Provided by publisher.
Identifiers: LCCN 2020036809 (print) | LCCN 2020036810 (ebook) | ISBN 9781644874394 (library binding) | ISBN 9781648341168 (ebook)
Subjects: LCSH: Negaprion brevirostris–Juvenile literature.
Classification: LCC QL638.95.C3 P485 2021 (print) | LCC QL638.95.C3 (ebook) | DDC 597.3/4-dc23
LC record available at https://lccn.loc.gov/2020036809
LC ebook record available at https://lccn.loc.gov/2020036810

Text copyright © 2021 by Bellwether Media, Inc. BLASTOFF! READERS and associated logos are trademarks and/or registered trademarks of Bellwether Media, Inc.

Editor: Rebecca Sabelko   Designer: Josh Brink

Printed in the United States of America, North Mankato, MN.

Table of Contents

What Are Lemon Sharks? 4
Lemons in the Sand 8
Group Hunters 12
Deep Dive on the Lemon Shark 20
Glossary 22
To Learn More 23
Index 24

What Are Lemon Sharks?

Lemon sharks are found throughout warm, coastal waters of the Atlantic and Pacific Oceans.

These sharks blend in to the sandy **bays** they call home. They swim in **coral reefs** and **mangrove** forests.

Lemon Shark Range

range =

The lemon shark population has fallen in recent years. They are now **near threatened**.

These sharks are overfished for their fins and skin. Human activity also destroys their **habitat**.

Lemons in the Sand

dorsal fins

Lemon sharks have thick bodies. They reach up to 12 feet (3.6 meters) long. They weigh up to 551 pounds (250 kilograms).

Lemon sharks have two **dorsal fins** that are about the same size.

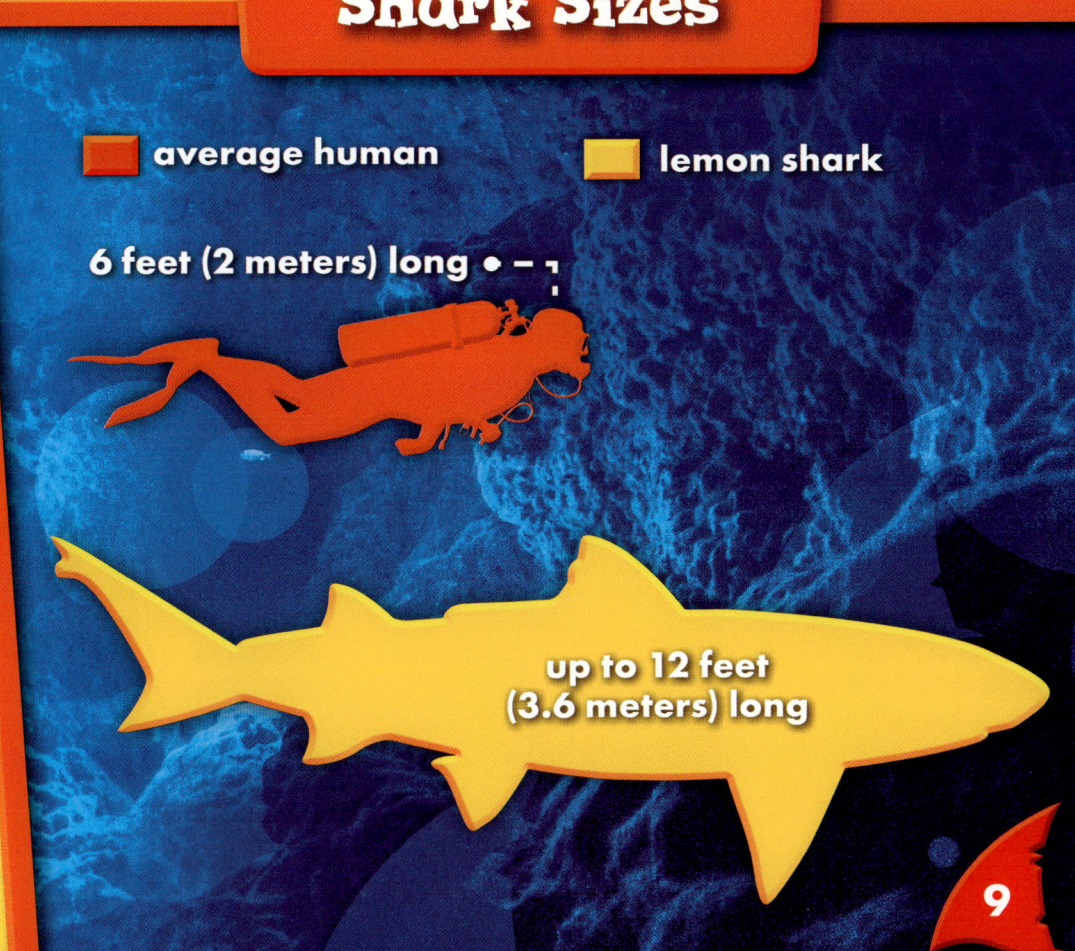

Shark Sizes

average human
lemon shark

6 feet (2 meters) long

up to 12 feet (3.6 meters) long

Lemon sharks have flat heads with round **snouts**. Their snouts have **sensors**. These help the sharks find food.

snout

Identify a Lemon Shark

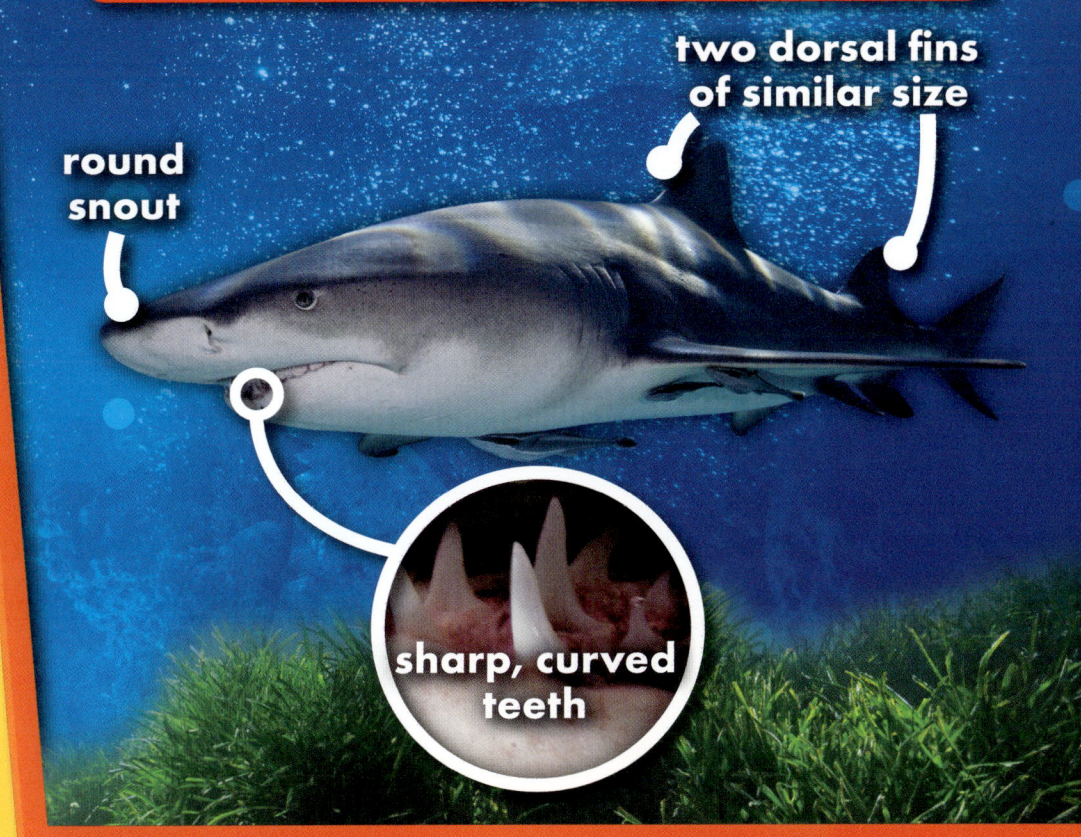

round snout

two dorsal fins of similar size

sharp, curved teeth

Sharp, curved teeth help these **predators** hold slippery fish.

11

Group Hunters

Lemon sharks often hunt in groups at dawn and dusk. Some groups have as many as 20 sharks!

Lemon sharks have a great sense of smell. It helps them find **prey**.

Lemon sharks eat prey whole. They shake their heads from side to side to tear the food apart!

They eat rays, crabs, and bony fish. They also eat sea birds and smaller sharks.

Lemon Shark Diet

bony fish

crabs

rays

Lemon sharks lie on the ocean floor during the day. They let small fish remove **parasites** from their skin.

They have to pump water over their **gills** to breathe. This uses a lot of **energy**.

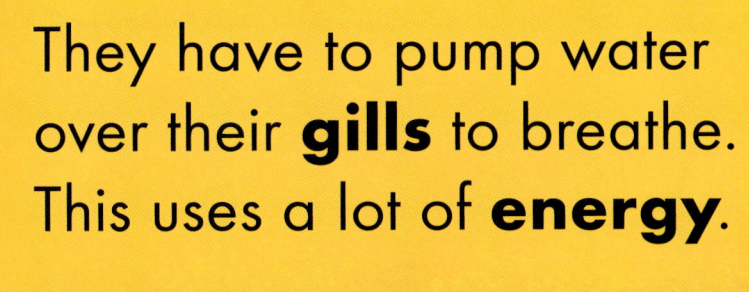

gills

Young lemon sharks may be eaten by larger sharks. But adults do not have natural enemies.

These sharks are important predators in the waters they call home!

Deep Dive on the Lemon Shark

two dorsal fins of similar size

LIFE SPAN:
up to **30 years**

LENGTH:
up to **12 feet (3.6 meters) long**

WEIGHT:
up to **551 pounds (250 kilograms)**

DEPTH RANGE:
0 to 300 feet (0 to 91 meters)

round snout

sharp, curved teeth

Least Concern	Near Threatened	Vulnerable	Endangered	Critically Endangered	Extinct in the Wild	Extinct

conservation status: near threatened

Glossary

bays—small areas filled with ocean water

coral reefs—structures made of coral that usually grow in shallow seawater

dorsal fins—fins at the top of a lemon shark's back

energy—the power to move and do things

gills—parts that help sharks breathe underwater

habitat—land with certain types of plants, animals, and weather

mangrove—related to groups of trees that grow in shallow saltwater or swamps

near threatened—may become extinct in the near future

parasites—living things that survive on or in other living things; parasites offer nothing for the food and protection they receive.

predators—animals that hunt other animals for food

prey—animals that are hunted by other animals for food

sensors—body parts that sense movement, heat, light, or sound

snouts—the noses of some animals

To Learn More

AT THE LIBRARY

Markle, Sandra. *What If You Could Sniff Like a Shark? Explore the Superpowers of Ocean Animals.* New York, N.Y.: Scholastic Press, 2020.

Meister, Cari. *Do You Really Want to Meet A Shark?* Mankato, Minn.: Amicus, 2016.

Pettiford, Rebecca. *Nurse Sharks.* Minneapolis, Minn.: Bellwether Media, 2021.

ON THE WEB

FACTSURFER

Factsurfer.com gives you a safe, fun way to find more information.

1. Go to www.factsurfer.com.

2. Enter "lemon sharks" into the search box and click 🔍.

3. Select your book cover to see a list of related content.

Index

Atlantic Ocean, 4
bays, 5
bodies, 8
breathe, 17
coral reefs, 5
deep dive, 20-21
dorsal fins, 8, 9, 11
enemies, 18
energy, 17
food, 10, 11, 14, 15
gills, 17
habitat, 7
heads, 10, 14
hunt, 12
mangrove forests, 5
ocean floor, 16
overfished, 7
Pacific Ocean, 4
parasites, 16
population, 6

predators, 11, 19
prey, 11, 12, 14, 15
range, 4, 5
sensors, 10
size, 8, 9
smell, 12
snouts, 10, 11
status, 6
swim, 5
teeth, 11
waters, 4, 17, 19
young, 18

The images in this book are reproduced through the courtesy of: Ian Scott, front cover (hero); Yann Hubert, p. 3; frantisekhojdysz, p. 4; Rodrigo Friscione/ Alamy, p. 6; Pat Bonish/ Alamy, p. 7; Luiz Felipe V. Puntel, p. 8; Michael Bogner, pp. 10, 23; June Jacobsen, p. 11; Damsea, p. 11 (sea floor); Greg Amptman, p. 11 (call out); NaluPhoto, p. 12; SeaTops/ Alamy, pp. 12-13; IS2010-09/ Alamy, p. 14; Peter Leahy, p. 15 (top left); Mirelle, p. 15 (top right); nicolavoisin44, p. 15 (bottom); Marion Kraschl, p. 16; Oceans Image/ Alamy, p. 17; Anita Kainrath, p. 18; Steve Bloom Images/ Alamy, p. 19; uwimages, pp. 20-21.